Math & Me

Math & Me

Embracing Success

Wendy Hageman Smith

&

B. Sidney Smith

Platonic Realms

Math & Me: Embracing Success

Platonic Realms
Appomattox, Virginia, USA.

Printed in the United States of America

10 9 8 7 6 5 4 3 2 1

ISBN 978-0-9636847-4-5 (pbk. : alk. paper)

An Authorized Instructor's Edition of this book with additional materials is available from PlatonicRealms.com.

For the many, many students
who have enriched our lives.

Contents

Introduction

IF you are holding this book in your hands, then you are probably a student. You might be in high school and looking towards college, or on your way to college, or perhaps already *in* college—and thinking about what's to come of it. And you are very likely considering—maybe with anticipation, maybe with dread—the math courses you will take (or are taking now!) and what they might do for you, for your plans, and for your career.

If that describes you, then congratulations! For you have come to the right place. This little book was written to help you make the math-y parts of your education work *for* you, to help you achieve your personal, academic, and professional goals. For whatever your goals may be, it is certain that an encounter with mathematics lies in your path. Some people find this prospect daunting, and many parts of this book were written precisely to meet the needs of students who anticipate a bad experience. But our main purpose is broader and grander. By presenting ideas and techniques harvested from the collective experience of thousands of students and teachers, this book will provide *every* student with the means to embrace success.

We start right off with everybody's favorite question: "Why do I need to know this?" Seriously, why study math at all? Even some academics have recently questioned the value of studying math beyond the basics, pointing to the power of our machines to take care of most tasks requiring the use of math. So we'll take an honest look at the pros and cons of studying math, consider its value, and help *you* decide if it's really worth it to you. (Spoiler alert: the smart money comes down on the side of studying at least *some* math in college, and more is usually better.)

Our next task will be to address the next most common question: *"What if I hate math?"* Not to worry, we've got you well covered on that issue. We'll begin by taking a constructive look at *math anxiety*, especially its causes and its effects. We'll untangle several common misconceptions people have about math, and especially about their own potential for succeeding at it, and help you understand the nature of the subject in a way you might not have previously. We will also outline several special techniques for disarming math anxiety that can benefit *every* student, not just the chronically math-anxious; for even those of us who love math can be overwhelmed by the stress of a high-stakes exam, or the difficulty of mastering a new topic.

We will then examine in detail the *actual* mechanics of succeeding in an *actual* math class. We'll look at special strategies for taking notes, and for getting the most out of lectures and other in-class experiences. We'll outline techniques for maximizing the value of both self-study and studying with others. The importance of homework is often misunderstood, so we'll pay special attention to making homework work *for* you and not *against* you. And finally, the big enchilada: Exams. We share our most valuable tips and tricks for getting the best scores possible, for not freaking out when you get stuck, and for letting the exam help you even after it's been graded.

Finally, we will demystify the college math curriculum by outlining the nature and purpose of the typical courses offered by most math departments. We'll consider which courses might be most valuable for *you*, given *your* goals, and we'll explore some reasons why you might eventually wish to go further with math than you expected to.

We notice we've written "we" a lot. Just who is "we"? Well, your authors are college math teachers. Have been a long time. Like many teachers we are fond of saying that "We will succeed when you do." Our purpose here is to give you the best tools we have to help you succeed at math—and so to succeed at anything and everything you want to achieve. Whether you are an up-and-coming scientist, musician, architect, entrepreneur, historian, artist, journalist, teacher, or even mathematician, in these pages you will find tools which—when carefully tucked away in your study-kit to be pulled out when needed—will make mathematics serve you well.

Happy hunting.

Chapter 1

Why Math?

WHAT parent or teacher has never heard these words: *Why do I have to learn this?* Whenever we hear the question we immediately give the correct answer: "You *don't* have to. You can (possibly) have a long and fulfilling life without ever learning any adult math."

There, we said it, and it's true.

But then, this isn't a question people ask when they are at their best, is it? This is a complaining sort of question, more intended to express frustration than to find wisdom. In our better moments we would likely ask a more sensible question: Why would I *want* to learn this? Now, that's a good question, and we have some answers for that too.

1.1   Math and Your World

To live in modern human society is to live in a world of mathematics. Some of it is right in our faces. *How many, how fast, how big,* and *how likely*

1

are just a few of the daily encounters we have with the quantitative side of life. But there is much more mathematics at work around us than is obvious.

Pick up an object, any object, manufactured in your own generation. How did it come to be? First, engineers and designers specified its physical characteristics, constituent materials, and proper use. Other technicians designed the manufacturing process. Business people implemented the financial and contractual structures that made its existence economically feasible, while accountants tracked the expenditures and revenues. Marketers and social scientists determined who would most benefit from its use, and how to disseminate information about it effectively. Computer scientists created the data and information processing systems that were needed at every step, from design to point of sale. And of course someone had to manage all these human resources.

Now let's look at the math. The need of some of these professionals for math is self-evident. The engineers are going to be using geometry, including analytic geometry, probably calculus, possibly even more advanced math, and the accountants will be crunching a lot of numbers.

But what about the visual designers, the ones who decide on such things as functionality, visual pattern, and even color? Their task is æsthetics, the discipline that concerns itself with what is beautiful and appropriate.

Æsthetics in every culture is intimately connected with that culture's geometrical insights, its intuitions about space and form, and its affinities to symmetry and pattern, rhyme and reason. All of these elements have a mathematical dimension.

Among the primary influences on western æsthetics are ancient Greek (Euclidian) geometry, the development of visual perspective (projective geometry) during the Renaissance, and most recently the fractal geometry and computational patterns that have become a distinct feature of the modern imagination. Good visual design may not often require the use of a calculator—but it is still nowhere without math.

Business-people rely on mathematical models to determine feasibility and profitability, models that require both calculus and linear algebra. The marketers and social scientists use both descriptive and inferential statistics, often using calculus and advanced computational methods. Computer scientists use formal logic, linear algebra, abstract algebra, and calculus. Even human resource managers must be able to cope effectively with quantitative information, in the interests of efficiency, productivity, and fairness.

And here we've only been talking about your hairbrush, your tennis shoes, or your favorite DVD. From the roads you drive on to your very system of government, every feature of modern civilization is made possible at least in part because of mathematics.

But wait, there's more; just because the world turns out to be somehow "mathematical" isn't the only reason for knowing about math. As a citizen and a human being, you have additional reasons for wanting a good grasp of what is often called *quantitative reasoning*.

Perhaps one of the best reasons is so that you can develop the habit of sound risk-assessment. In our life we face many hazards; from crime, from

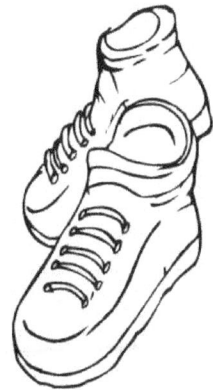

disease and disability, from personal accidents, and from acts of nature like storms and earthquakes, to name a few. Whether we fall victim to one of these hazards isn't something we can always control directly, because even the most careful person can't prevent the unforeseeable. What we *can* do is learn to assess relative risk, and people who are good at doing so generally live longer and are both healthier and happier. However, assessing risk correctly requires that we be able to reason well—especially about probability and statistics.

Consider the use of alcohol, for instance. Some studies have indicated that moderate use of alcohol promotes good health. Other studies have directly linked the use of alcohol to cancer, heart-disease, and mental-health problems. The findings of such studies are always expressed as *statistical* findings. Without any understanding of what the science of statistics is or how it is used, you cannot hope to use the information provided to make intelligent choices about the use of alcohol.

What family planning methods are appropriate for you? How much insurance do you need? Should you give your children vitamins? How much sun should you get? Exercise? Sleep? Media is full of answers to these questions, but only with good risk-assessment tools can you choose wisely among those answers.

Finally, every adult person has a role to play as a citizen. In a free society, the present well-being and prospects of all people depend on the choices made by its citizens.

For example, should you vote? Many people do not, often because they think (or hope) that their vote is of little importance. Mathematics has much to teach us about voting (it's called "social-choice theory"), and about what voting systems can and cannot achieve.

Supposing you do vote, many of the questions you must decide as a voter are quantitative. For instance, there is an ongoing debate in the United States over whether the use of statistical methods would be more accurate in taking the Constitutionally-mandated population census every ten years than the direct-counting "enumeration" that has been used up to now. Or again, much debate is spent on the deficit; but what does one-trillion dollars really mean in human terms? How much food safety regulation is enough to ensure the public health, without placing a burden on manufacturers that causes greater harm to public well-being? What fuel-efficiency standards are appropriate? What are "marginal" tax rates? How is the unemployment rate calculated? What is the relationship between interest rates and inflation?

Indeed, there are very few policy questions voters are asked to consider that do *not* call for the use of quantitative reasoning to understand them correctly.

The conclusion here is inescapable. Just like literacy, the capacity to use mathematics where appropriate opens doors that before were locked or hidden. Math allows us to see the world more clearly, to understand it more deeply, and to engage it more effectively.

It's good to know.

1.2 Math and Your Career

As a college student, one of your foremost concerns is your choice of major and (presumably) of your subsequent career. Many students come to college with definite interests, some even with their careers already mapped out in their minds. But most students find during their first year or two in college that there is a far greater range of possibilities than they could possibly have guessed. It is a time of great personal growth as well, and no one can predict with certainly the subject of study that will set a student on fire.

At the least, math should not be an obstacle. Sadly, many students prejudice themselves against majors that might otherwise interest them, solely because they are afraid of the required coursework in mathematics. We hope that by working through this book, taking possession of any math anxiety you may have, and learning the best strategies for succeeding in your math courses, you will open up new pathways to your future success.

More often than ever before, college students today report that one of their biggest concerns is financial security and professional success. It is worth reviewing the income requirements that correspond to your life plans.

cost of living

For instance, a person living alone requires an income of \$20,000 to \$35,000, depending on where they live, just to maintain a lifestyle that includes a small apartment, a car, and minimal expenses. (So, not including travel, a hobby, or savings.) A family of four with one parent working and one staying home needs an income of \$45,000 to \$75,000 for a similarly no-frills lifestyle. If both parents work, the minimum income goes up to $60,000 to $90,000 because of the second earner's transportation and other work-related expenses, plus the cost of childcare.

Remember, these figures are for a very modest life, without saving for children's college or contributing to private retirement investments. To include retirement and/or emergency savings, family vacations, and college for two children, in most areas of the country an income of $100,000 per year is about the minimum household income needed to raise a family. This is contrary to many people's expectations because it used to be thought that a 6-figure income meant being quite comfortable. That expectation was meaningful 30 years ago, but consider: to have the same spending power as someone who had an income of $100,000 in 1980, you would need to make more than $250,000 in 2013.

If you have concerns about your income potential, consider *Jobs Rated 2013*, a ranking of 200 jobs/careers ranked from best to worst by the online job-search service CareerCast.com. Their rankings took into account not only average salary but also job environment, job-related stress, and hiring prospects. The 20 top-ranked jobs were:

high-ranked jobs

1. Actuary

2. Biomedical Engineer

3. Software Engineer

4. Audiologist

5. Financial Planner

6. Dental Hygienist

7. Occupational Therapist

8. Optometrist

9. Physical Therapist

10. Computer Systems Analyst

11. Chiropractor

12. Speech Pathologist

13. Physiologist

14. University Professor

15. Veterinarian

16. Dietitian

17. Pharmacist

18. Mathematician

19. Sociologist

20. Statistician

Three of these jobs, statistician, mathematician, and actuary (the number 1 job), actually require a degree in mathematics, but only two of these jobs (dental hygienist and chiropractor) have no specific math requirements beyond college algebra. Every other life-sciences or medical-related career requires a firm grasp of statistics, and most require calculus. The computer-related fields such as systems analyst and software engineer also require statistics and calculus, and usually more advanced courses. In short, you don't get on this list without *some* college math, and more than half of these require quite a lot of math. It is also notable that the jobs on this list that require the least math also pay the least; the average hygienist salary is $54,000, while the average mathematician's or actuary's salary is about twice that (around $100,000), as are the salaries for those in computer-related fields.

On the other hand, none of the twenty worst-rated jobs, like waiter/waitress, personal care aide, and manicurist, requires mathematics beyond the high school level. Unsurprisingly, these are also the lowest-paid jobs, around $20,000. The conclusion to be drawn is clear: graduates with college training in mathematics generally qualify for a much higher economic and social status than those who don't. As a report by CNN Money on the most lucrative college degrees put it, "Math is at the crux of who gets paid." This is hardly surprising, since the historical development of our advanced civilization paralleled the growth in our mathematical knowledge, knowledge that allows us to build powerful models of our world and to solve problems that were once beyond human reach.

low-ranked jobs

A number of editorials and essays have appeared in recent years challenging the expectation that everyone should have to study mathematics at least through college algebra. These commentators—some of them even professional academics—argue that learning mathematical skills is no longer relevant to most people's lives, and that it is simply a waste of their time to make them do it. Let us leave aside for the moment the dispiriting assumption that only "useful" things are worthy of study, and just notice that these articles have been written by people who have already achieved a measure of professional success. For the student struggling with the choices that will ultimately determine their own success, it is clear that dismissing math out of hand is a poor strategy.

But unless you were planning on one of the above jobs, you are probably thinking right now, "what about the career *I'm* interested in?" Many are drawn to the liberal arts, to subjects of great cultural and social importance that are not commonly thought of as mathematical. If this includes you, it is entirely reasonable to ask, "why bother with math?" Well, suppose you want to be a/an...

Artist.

It is of course not necessary to study math to be an artist. It is only necessary to study math to be a *great* artist. For consider, the western tradition of art was born in the same place, and at the same time, as mathematics itself, and this is no accident. It was the ancient Greeks who gave us our first clear ideals of form, proportion, structure, and beauty, and for them "the arts" included not only sculpture and painting, but mathematics too. They considered them inseparable.

Mighty is geometry; joined with art, resistless.

—Euripides

After the decline of Hellenic civilization, art and mathematics languished. During this period Islamic civilization made enormous and important contributions, but there were comparatively few mathematical or artistic advances in Europe. When art and math were at last reborn in the West they were—again—reborn together. The Renaissance rediscovered the Hellenic principles of form, taking Western art from the flat, dimensionless, iconographic images of the Dark Ages to the naturalism, grandeur, and dimensionality that characterizes the work of the great masters. And those same great masters were no strangers to the revolutions in mathematics that were also occurring; indeed they saw it as crucial to their art. What artists call "perspective" and mathematicians call "projective geometry" was discovered not just simultaneously by artists and mathematicians, but together by artists who *were* mathematicians, mathematicians who were artists.

Geometry is the right foundation of all painting.

—Albrecht Dürer

In the modern era mathematics continued to have a strong influence on art. Just as ancient and renaissance artists used the Golden Ratio and other principles from classical mathematics in their compositions, so modern artists began to incorporate ideas such as algebraic symmetry, mathematical infinity, and non-Euclidean geometry into their work. Pablo Picasso and Salvador Dali, for instance, took direct inspiration from such mathematical themes. Some have even used mathematical ideas themselves as the subject of their work, most famously the Dutch artist M.C. Escher.

Whoever despises the high wisdom of mathematics nourishes himself on delusion.

—Leonardo da Vinci

Contemporary artists have yet another reason to acquire mathematical literacy: the computer. Digital technologies are now central to many artists' work, not just in graphic arts but in the plastic arts as well. Understanding such things as color-depth and computational complexity are an important foundation for working with these technologies, and "getting" the math behind it all is the way to acquire that foundation.

As college math educators we can testify that some of our most successful students have been dual majors; art *and* math. It is remarkable how the two activities—we dare say the two ways of thinking—complement and support one another.

Musician.

Like painting and sculpture, music is deeply interwoven with mathematics from at least the time of Pythagoras (c. 500 BCE), whose followers gave us the 12-tone chromatic scale still used in music today, a scale based on numerical ratios.

There is geometry in the humming of the strings.

—Pythagoras

Not only the musical scale but keys, intervals, harmony, meter, and rythm all have a mathematical dimension that is central to understanding them fully. Composers use numerical relationships in selecting musical form. Pitch is a measure of the frequency of sound itself, and tones used in music may be analyzed in terms of the mathematical properties of those frequencies. Modern abstract algebra may be used to analyze the relationships among the whole and half-steps that define the various types of musical keys (what the ancient Greeks called "modes") such as the major, minor, pentatonic, and blues keys.

May not music be described as the mathematics of sense, mathematics as music of the reason? The musician feels mathematics, the mathematician thinks music – music the dream, mathematics the working life.

—James Joseph Sylvester

Analyses of some of the greatest composers' works have revealed startling mathematical connections. Mozart's compositions are being studied for their employment of mathematical principles of symmetry. The golden ratio commonly appears in many famous compositions, such as Debussy's *Image; Reflections in Water*. Among the composers who are known to have deliberately incorporated the golden ratio and other mathematical forms into their compositions are Béla Bartók and Erik Satie.

Although most musicians, perhaps, are not also mathematicians, it is actually very common for people to do both. Some professional mathematicians have even made a name for themselves as musicians, such as the musical humorist, composer, and performer Tom Lehrer. Like the other fine arts, the study of music is enhanced and complemented by the study of mathematics, and *vice versa*.

Writer.

There may seem to be no two academic majors farther apart than English literature/composition and mathematics. And yet some of the finest writers have also been mathematicians; a facility with one seems always to foretell a facility with the other. As Alexander Nazaryan wrote in The New Yorker (2012),

> What ballet is to football players, mathematics is to writers, a discipline so beguiling and foreign, so close to a taboo, that it actually attracts a few intrepid souls by virtue of its impregnability. The few writers who have ventured headlong into high-level mathematics—Lewis Carroll, Thomas Pynchon, David Foster Wallace—have been among our most inventive in both the sentences they construct and the stories they create.

Other writers (among hundreds) who have incorporated mathematical ideas directly into their work include the Argentinian poet Jorgé Louis Borges; the novelists Samuel Beckett, John Updike, Charles Dickens, and Leo Tolstoy; short-story writers John Cheever and O. Henry; playwright George Bernard Shaw; the mystery writers Sir Arther Co-

A man should be learned in several sciences, and should have a reasonable, philosophical and in some measure a mathematical head, to be a complete and excellent poet.

—John Dryden

nan Doyle and Agatha Christie; the children's author Jules Feiffer; screenwriter and comedian Tina Fey; satirist Aldous Huxley; horror writers H.P. Lovecraft and Mary Wollstonecraft Shelley; and of course dozens of science-fiction writers such as Piers Anthony, Isaac Asimov, Robert Heinlein, and (mathematician) Rudy Rucker.

To the extent that writers explore the human world, and to the extent that the human world is rife with mathematical relations, connections, and ideas, it makes a great deal of sense for an aspiring writer to set some exploration of mathematics itself as an important professional goal.

If a man is at once acquainted with the geometric foundation of things and with their festal splendor, his poetry is exact and his arithmetic musical.

—Ralph Waldo Emerson

Philosopher

As with the arts, in the West mathematics and philosophy have their roots in the contributions of the ancient Greeks, to whom the notion that they were somehow *separate* realms of knowledge would have been incomprehensible. And of course they are not separate, but deeply intertwined.

Most philosophy programs require a course in formal logic—a topic which, even if it is taught in the philosophy department, is still part of mathematics. Those who study metaphysics benefit from an acquaintance with the mathematical field of analysis (which begins with calculus) for its construction of the continuum and—relatedly—its solutions of the ancient problems of Zeno. Those concerned with ontology benefit from knowing set theory; epistemology can be illuminated by model theory and formal languages.

Every good mathematician is at least half a philosopher, and every good philosopher is at least half a mathematician.

—Gottlob Frege

Unsurprisingly, the *philosophy of mathematics* is itself a thriving discipline, with important international journals and conferences.

Historian

It is an interesting fact that among all the primary artifacts of human culture, things like language, music, and art, mathematics alone is traceable in the historical record from its earliest beginnings. The historian will find in mathematics, in its development and elaborations—and above all in its contributions—a diagram of the human saga stretching over 10,000 years.

History from the classical period to the present is especially remarkable for the mileposts provided by mathematics. The importation of Arabic numerals and arithmetic in the 11th century stimulated revolutions in commerce, the birth of European trade empires, and the invention of banking; the rediscovery of classical mathematics—together with the new techniques of algebra learned from Islamic scholars—led to the birth of theoretical science; the discovery of the calculus gave us physics; the discovery of statistics opened the doors of modern experimental science; and the work of an obscure Cambridge mathematician in the 1930's issued forth in the general (programmable) computer. From the ancient Egyptians, marking out flooded fields with Pythagorean right triangles, to the modern bank securing its electronic transactions with advanced techniques from number theory, the human world has progressed along a track that is told in the story of mathematics.

In these days of conflict between ancient and modern studies, there must surely be something to be said for a study which did not begin with Pythagoras and will not end with Einstein, but is the oldest and youngest of all.

—Godfrey Harold Hardy

15

Since every period in human history is shaped in part by its mathematics, the historian has a natural interest in knowing a little mathematics herself.

* * * * *

Mathematics is a curious subject. It isn't really *about* anything, except itself. It sometimes isn't needed for ordinary things, but it is an ingredient—often the key ingredient—in every extraordinary thing. Because at its heart it is the careful exploration of every kind of structure we can imagine, there is no amount of studying math that doesn't reward the student with new capacities, and above all with new potentials for imaginative thinking. Whatever your career plans, wherever you see yourself in 10, 20, or 40 years, mathematics can only give your plans a lift.

1.3 Response: Math Connections

1. Using an internet search engine or an online resource such as payscale.com, livecareer.com, or the US Dept. of Labor's Bureau of Labor Statistics website, review the typical salaries for careers you might consider.

2. Write down three careers you believe you would find rewarding, *and* that are likely to meet your income goals. For each of these, list two or three ways you think math might be used, or in which it might contribute to your success.

3. Pick your favorite academic subject (besides math), and then run an internet search using

the name of that subject together with "mathematics." Explore the result pages, and make
a note of anything you learn about the connections between your favorite subject and math.

Chapter 2

Math Anxiety

As you can see from the old rhyme at right, dislike of mathematics is neither new nor uncommon. Math is a fundamental artifact of human culture and a core element of both academic and societal advancement, and yet—almost uniquely—it is the focus of deeply negative feelings in a great many people. On an opinion gathering website some 86% of respondents to a question about algebra indicated that they "hate" it. Together with geometry, it was identified as the most disliked of all school subjects.

Multiplication is vexation,
Division is just as bad;
The Rule of Three perplexes me,
And Practice drives me mad.

—Old Rhyme

If you are reading this, it is not at all unlikely that you have some negative feelings about math yourself. This chapter is for you.

2.1 What Is Math Anxiety?

A famous stage actress was once asked if she had ever suffered from stage fright, and if so how she had gotten over it. She laughed at the interviewer's naïve assumption that, since she was an accomplished actress *now*, she must not feel that kind of anxiety. She assured him that she had *always* had stage fright, and that she had *never* gotten over it. Instead, she had learned to walk on stage and perform—in spite of it.

Like stage fright, math anxiety can be a disabling condition, causing humiliation, resentment, and even panic. Consider these testimonials from students:

math anxiety
testimonials

- When I look at a math problem, my mind goes completely blank. I feel stupid, and I can't remember how to do even the simplest things.

- I've hated math ever since I was nine years old, when my father grounded me for a week because I couldn't learn my multiplication tables.

- In math there's always one right answer, and if you can't find it you've failed. That makes me crazy.

- Math exams terrify me. My palms get sweaty, I breathe too fast, and often I can't even make my eyes focus on the paper. It's worse if I look around, because I'll see everybody else working, and know that I'm the only one who can't do it.

- I've never been successful in any math class I've ever taken. I never understand what the teacher is saying, so my mind just wanders.

- Some people can do math—not me!

All of these students are expressing math anxiety: a feeling of intense frustration or helplessness about mathematics. What students usually don't realize is that these feelings about math are very common. Even successful mathematicians, like the actress mentioned above, can be prone to anxiety—even about the very thing they do best and love most.

2.2 Social and Educational Roots

Imagine that you are at a dinner party, seated with many people at a large table. In the course of conversation the person sitting across from you laughingly remarks, "Of course, I'm illiterate!" What would you say? Would you laugh along with him or her and confess that you never really learned to read either? Would you expect other people at the table to do so?

Now imagine the same scene, only the guest across from you says, "Of course, I've never been any good at math!" What happens this time? Naturally, you can expect other people at the table to chime in cheerfully with their own claims to having "never been good at math"—the implicit message being that no ordinary person is.

It's a fact that mathematics has a tarnished reputation in our society. It is commonly accepted that math is difficult, obscure, and of interest only to "nerds" and "geeks"—not a flattering characterization. Consequently the study of math carries with it a stigma, and people who are talented at math or profess enjoyment of it are often treated as though they are not quite normal. Alarmingly,

Poor teaching leads to the inevitable idea that the subject (mathematics) is only adapted to peculiar minds, when it is the one universal science, and the one whose ground rules are taught us almost in infancy and reappear in the motions of the universe.

—H.J.S. Smith

many school teachers—even those whose job it is to teach mathematics—communicate this attitude to their students directly or indirectly, so that young people are often exposed to an anti-math bias at an impressionable age.

It comes as a surprise to many people to learn that this attitude is not shared by other societies. In Russian or German culture, for example, mathematics is viewed as an essential part of literacy, and an educated person would be chagrined to confess ignorance of basic mathematics. (It is no accident that both of these countries enjoy a centuries-long tradition of leadership in mathematics.)

Jaundiced attitudes toward mathematics are worsened by the way it is sometimes taught. In the past, teachers were often trained to believe that students should learn new skills by rote, that is, by memorization and repetition. In mathematics, this meant that a particular type of problem was presented, together with a technique of solution, and these were practiced until sufficiently mastered. The student was then hustled along to the next type of problem, with its technique of solution, and so on. This method of teaching mathematics has been compared to being invited to the most wonderful restaurant in the world—and then being forced to eat the menu! Little wonder that the learning of mathematics seems to many people a dull and unrewarding enterprise, when the very meat of the subject is boiled down to the gristle before it is served.

There is both good news and bad news about the direction mathematics education is now taking. The good news is that reform efforts in the teaching of mathematics have been under way for

Students must learn that mathematics is the most human of endeavors. Flesh and blood representatives of their own species engaged in a centuries long creative struggle to uncover and to erect this magnificent edifice. And the struggle goes on today. On the very campuses where mathematics is presented and received as an inhuman discipline, cold and dead, new mathematics is created. As sure as the tides.

—J.D. Phillips

several years, and many—if not all—teachers of mathematics have adopted methods based on constructivist or other progressive models of learning. Good teaching empowers students to discover the range and beauty of mathematical ideas, free of the stigmas engendered by social and educational bias. The bad news is that the flexibility and independence that teachers once enjoyed, and that allowed them to implement the most appropriate teaching strategies for their students, have been sharply curtailed by the trend towards more and more standardized testing. When teachers have to teach to a test, comprehension and discovery learning get pushed aside in favor of rote responses to formulaic questions.

The Mind is not a vessel to be filled. It is a fire to be kindled.

—Plutarch

Finally, young women sometimes face additional barriers to success in mathematics. Remarkably, even now in the 21st century, school age girls are often discouraged by parents, peers, and teachers with the admonition that mathematics "just isn't something girls do." One student recalled approaching her junior high school geometry teacher after class with a question about what the class was studying. "You don't need to know about that stuff," he told her, smiling condescendingly. (And, needless to say, he didn't answer her question.) Rank sexism such as this is only part of the problem. Female students still report that friends, family members, and even their junior and senior high school instructors impressed upon them the undesirability of pursuing the study of mathematics. For all adolescents there is concern about how one is viewed by members of the opposite sex, and being a "geek" is not seen as the best strategy, especially for girls. Peer pressure is the mortar in that wall.

And parents, often even without knowing it, can facilitate this anxiety and help to discourage their daughters from maintaining an open mind and a natural curiosity towards the study of science and math.

Together these social and educational factors lay the groundwork for many widely believed myths and misconceptions about the study of mathematics. We will take a close look at these next.

2.3 Response: Questionnaire

This questionnaire was developed as a research tool in mathematics education, to help future teachers to understand and identify common signs of math anxiety. As a student, you can use it to explore your own feelings and attitudes about math. Respond to each statement by placing a number in the space provided, as follows:

Strongly disagree: 1

Somewhat disagree: 2

No Opinion: 3

Agree somewhat: 4

Agree strongly: 5

Don't read too much into them—just go with your first reaction.

1. ____ I believe mathematics is difficult to learn unless you have the brain for it.

2. ____ When in a mathematics classroom, I am reluctant to ask questions.

3. _____ With mathematics, there is always one interpretation and one right answer.

4. _____ I find I have problems using mathematical concepts when I need to.

5. _____ I have little to contribute in a mathematics classroom.

6. _____ Most mathematics I learn in the classroom is relevant to my life.

7. _____ I like math.

8. _____ Knowing about math is important for really understanding how the world works.

9. _____ Mathematics is usually easy for me.

10. _____ I do better in other classes than I do in math classes.

11. _____ Math is probably not something I will ever use after college.

12. _____ Math is something I expect to use often in daily life.

13. _____ I generally feel competent to handle ordinary math problems.

14. _____ Men are generally better at math than women are.

15. _____ After arithmetic, math is pretty much a subject for specialists. Ordinary people can't understand it.

16. _____ An understanding of mathematics is important to be an informed citizen.

17. _____ It's hard for me to understand why anyone would want to spend his or her life studying mathematics.

18. _____ There is little reason for artistic people to be interested in mathematics.

19. _____ Math is a dead subject—cut and dried.

20. _____ Understanding mathematics requires creative insight.

21. _____ Mathematics is primarily a practical endeavor, best suited to practical people, like accountants.

Once you have put a response for each statement, put a check mark in front of the statements that seem to express confidence about mathematics and its value in daily life. Did you generally agree with these statements? Why or why not? Discuss.

2.4 Math Myths

A host of common but erroneous ideas about mathematics are available to the student who suffers math anxiety. These have the effect of justifying or rationalizing the fear and frustration he or she feels, and when these myths are challenged a student may feel defensive. This is quite natural. However, it must be recognized that loathing of mathematics is an emotional response, and the first step in overcoming it is to appraise one's opinions about math in a spirit of detachment. Consider the five most prevalent math myths, and see what you make of them.

Myth 1: Aptitude for Math Is Inborn

This belief is the most natural in the world. After all, some people just are more talented at some things (music and athletics come to mind) and to some degree it seems that these talents must be inborn. Indeed, as in any other field of human endeavor, mathematics has had its share of prodigies. Karl Gauss helped his father with bookkeeping as a small child, and the Indian mathematician Ramanujan discovered deep results in mathematics with little formal training. It is easy for students to believe that doing math requires a *math brain*, one in particular which they have not got.

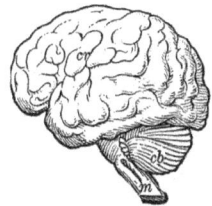

Math Brain

But consider: to generalize from "three spoons, three rocks, three flowers" to the number "three" is an extraordinary feat of abstraction, yet every one of us accomplished this when we were mere toddlers! Mathematics is indeed inborn, but it is inborn in all of us. It is a human trait, shared by the entire race. Reasoning with abstract ideas is the province of every child, every woman, every man. Having a special genetic makeup is no more necessary for success in this activity than being Mozart is necessary to humming a tune.

Ask your math teacher or professor if he or she became a mathematician in consequence of having a special brain. (Be sure to keep a straight face when you do this.) Almost certainly, after the laughter has subsided, it will turn out that a parent or a teacher was responsible for helping your instructor discover the beauty in mathematics, and the rewards it holds for the student—and decidedly not a special brain.

Myth 2: To Be Good At Math You Have To Be Good At Calculating

Some people count on their fingers. Invariably, they feel somewhat ashamed about it, and try to do it furtively. But this is ridiculous. Why shouldn't you count on your fingers? What else is a Chinese abacus, but a sophisticated version of counting on your fingers? Yet people accomplished at using the abacus can outperform anyone who calculates figures mentally.

Modern mathematics is a science of ideas, not an exercise in calculation. It is a standing joke that mathematicians can't do arithmetic reliably, and there is a serious message in this: being a wiz at figures is not the mark of success in mathematics.

This bears emphasis. A pocket calculator has no knowledge, no insight, no understanding, yet it is better at addition and subtraction than any human will ever be. And who would prefer being a pocket calculator to being human?

This myth is largely due to the methods of teaching discussed in Chapter 1, which emphasize finding solutions by rote. Indeed, many people suppose that a professional mathematician's research involves something like doing long division to more and more decimal places, an image that makes mathematicians smile sadly. New mathematical ideas—the object of research—are precisely that. Ideas. And ideas are something we can all relate to. That's what makes us people to begin with.

Myth 3: Math Requires Logic, Not Creativity

The grain of truth in this myth is that, of course, math does require logic. But what does this mean? It means that we want things to make sense. We don't want our equations to assert that 1 is equal to 2. This is no different from any other field of human endeavor, in which we want our results and propositions to be meaningful—and they can't be meaningful if they do not jive with the principles of logic that are common to all humankind. Mathematics is somewhat unique in that it has elevated ordinary logic almost to the level of an art form, but this is because logic itself is a kind of structure—an idea—and mathematics is concerned with precisely that sort of thing.

But it is simply a mistake to suppose that logic is what mathematics is about, or that being a mathematician means being uncreative or unintuitive, for exactly the opposite is the case. The great mathematicians, indeed, are poets in their soul.

How can we best illustrate this? Consider the ancient Greeks, such as Thales, Pythagoras, and Euclid, who first brought mathematics to the level of an abstract study of ideas. They noticed something truly astounding: that the musical tones most pleasing to the ear are those achieved by dividing a plucked string into ratios of integers. For instance, the musical interval of a "fifth" is achieved by plucking a taut string whilst pressing the finger against it at a distance exactly three-fourths along its total length. From such insights, the Pythagoreans developed an elaborate and beautiful theory of the nature of physical reality, one based on number. And to them we owe an immense debt,

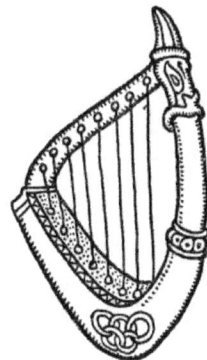

for to whom does not music bring joy? Yet no one could argue that music is a cold, unfeeling enterprise of mere logic and calculation.

If you remain unconvinced, use the internet to look up and examine the works of Dutch artist M.C. Escher. You will find the creative legacy of an artist with no advanced training in math, but whose works celebrate mathematical ideas in a way that slips them across the transom of our self-conscious anxiety, presenting them afresh to our wondering eyes.

Myth 4: In Math What Counts Is Getting The Right Answer

If you are building a bridge, getting the right answer counts for a lot, no doubt. Nobody wants a bridge that tumbles down during rush hour because someone forgot to carry the 2 in the 10's place! But are you building bridges, or studying mathematics? Even if you are studying math so that you *can* build bridges, what matters right now is understanding the concepts that allow bridges to hang magically in the air—not whether you always remember to carry the 2.

That you be methodical and complete in your work is important to your math instructor, and it should be important to you as well. This is just a matter of doing what you are doing as well as you can do it—good mental and moral hygiene for any activity. But if any instructor has given you the notion that "the right answer" is what counts most, put it out of your head at once. Nobody overly fussy about how his or her bootlace is tied will ever stroll at ease through the landscapes of mathematics.

Myth 5: Men Are Better Than Women At Mathematics

If there is even a ghost of a remnant of a suspicion in your mind about gender making a whit's difference in students' mathematics aptitude, slay the beast at once. Special vigilance is required when it comes to this myth, because it can find insidious ways to affect one's attitude without ever drawing attention to itself. For instance, female students sometimes express the view that—although of course they do not believe in a gender gap when it comes to *ability*—still it seems to them a little *unfeminine* to be good at math. There is no basis for such a belief, and in fact a sociological study several years ago found that female mathematicians are, on average, slightly *more* feminine than their non-mathematician counterparts.

Sadly, the legacy of generations of gender bias, like our legacy of racial bias, continues to shade many people's outlooks, often without their even being aware of it. It is every student's, parent's, and educator's duty to be on the lookout for this error of thought, and to combat it with reason and understanding wherever and however it may surface.

Across the centuries, from Hypatia to Amalie Nöther to thousands of contemporary women in school and university math departments around the globe, female mathematicians have been and remain full partners in creating the rich tapestry of mathematics. Many outstanding web sites pro-

Hypatia of Alexandria

vide information about historical and contemporary women in mathematics.

2.5 Thinking About Math Myths

Write a one, two, or three-sentence response to each of these questions.

1. Of the five math myths discussed above, which one do you find the most plausible? Which is the least plausible?

2. Which math myth do you think is most likely to be believed by people you know, and why?

3. Another commonly believed myth is that people are "right-brained" or "left-brained" in the same way that they are right-handed or left-handed. Given that this is another way of rationalizing frustration, what would be an appropriate way to help students who believe they are "right-brained" to achieve their educational goals?

4. Many of the greatest artists and mathematicians through the centuries have insisted that the creative impulses in each discipline are much the same. In what ways could art be mathematical, or math be artistic?

Chapter 3

Coping with Math Anxiety

EVEN though all of us suffer from math anxiety to some degree—just as anyone feels at least a little nervous when speaking to an audience—for some of us it is a serious problem, a burden that interferes with our lives, preventing us from achieving our goals. The first step, and the one without which no further progress is possible, is to recognize that math anxiety is an emotional response.

As with any strong emotional reaction, there are constructive and unconstructive ways to manage math anxiety. In this chapter we explore these in detail, and provide time-tested techniques for getting math anxiety under control.

3.1 Rationalizing, Suppressing, Denying

rationalization

Unconstructive (and even damaging) ways of coping include rationalization, suppression, and denial. By "rationalization," we mean finding reasons why it is okay and perhaps even inevitable—and therefore justified—for you to have this reaction. The Math Myths discussed in Chapter 2 are examples of rationalizations, and while they may make you feel better (or at least less bad) about having math anxiety, they will do nothing to lessen it or to help you get it under control. Therefore, rationalization is unconstructive.

suppression

By "suppression" is meant having awareness of the anxiety—but trying very, very hard not to feel it. This is very commonly attempted by students, and it is usually accompanied by some pretty severe self-criticism. Students believe that they shouldn't feel this anxiety, that it's a weakness that they should overcome, by brute force if necessary. When this effort doesn't succeed (as invariably it doesn't) the self-criticism becomes ever harsher, leading to a deep sense of frustration and often a severe loss of self-esteem—particularly if the stakes for a student are high, as when his or her career or personal goals are riding on a successful outcome in a math class, or when parental disapproval is a factor. Consequently, suppression of math anxiety is not only unconstructive, but can actually be damaging.

denial

Finally, there is denial. People using this approach carefully construct their lives so as to avoid all mathematics as much as possible. They choose college majors, and later careers, that don't require any math, and let the bank or their spouse balance the checkbook. This approach has the advantage that feelings of frustration and anxiety about math are mostly avoided. However, their lives are drastically constrained, for in our society fewer than 25% of all careers are, so-to-speak, "math free," and thus their choices of personal and professional goals are severely limited. As we saw in Chapter 1, most of these math-free jobs are also low-status and low-pay.

People in denial about mathematics miss out on something else too, for the student of mathematics learns to see aspects of the structure and beauty of our world that can be seen in no other way, and to which the "innumerate" necessarily remain forever blind. It would be a lot like never hearing music, or never seeing colors. (Of course some people have these disabilities, but being in denial about mathematics is a condition you can change.)

3.2 Taking Possession of Math Anxiety

Okay, so what is the constructive way to manage math anxiety? We'll call it "taking possession." It involves making as conscious as possible the sources of math anxiety in one's own life, accepting those feelings without self-criticism, and then learning strategies for disarming math anxiety's influence on one's future study of mathematics. (These strategies are explored in depth in the next section.)

Begin by understanding that your feelings of math anxiety are not uncommon, and that they def-

taking possession

your math history

initely do not indicate that there is anything wrong with you or inferior about your ability to learn math. For some this can be hard to accept, but it is worth trying to accept—since after all it happens to be true. This can be made easier by exploring your own "math history." Think back across your career as a math student, and identify those experiences which have contributed most to your feelings of frustration about math. For some this will be a memory of a humiliating experience in school, such as being made to stand at the blackboard and embarrassed in front of one's peers. For others it may involve interaction with a parent. Whatever the principle episodes are, recall them as vividly as you are able to. Then, write them down. This is important. After you have written the episode on a sheet(s) of paper, write down your reaction to the episode, both at the time and how it makes you feel to recall it now. (Do this for each episode if there is more than one.)

After you have completed this exercise, take a fresh sheet of paper and try to sum up in a few words what your feelings about math are at this point in your life, together with the reason or reasons you wish to succeed at math. This too is important. Not until after we lay out for ourselves in a conscious and deliberate way what our feelings and desires are towards mathematics will it become possible to take possession of our feelings of math anxiety and become free to implement strategies for coping with those feelings.

talking it out

It can be enormously helpful to share your memories, feelings, and goals with others. This process of dialogue and sharing—though it may seem just a bit on the goopy side—invariably brings out of each student his or her own barriers to math, often helping them become completely conscious of these barriers for the first time. Just as important, it helps all students understand that the negative experiences they have had, and their reactions to them, are shared one way or another by almost everyone.

If you do not have the opportunity to engage in a group discussion in a classroom setting, find friends or relatives whom you trust to respect your feelings, and induce them to talk about their own experiences of math anxiety and to listen to yours.

Once you have taken possession of your math anxiety in this way, you will be ready to implement the strategies outlined in the next chapter.

3.3 Response: Discussion and Writing

In addition to the group discussion and written responses outlined in the previous section, your instructor may ask you to complete a two to three-page paper. This should be an exploratory essay in which you reflect upon and respond in full to the following questions:

1. Which experiences in your own history as a math student have most contributed to any negative feelings you may have about math?

2. How similar are your own experiences and feelings concerning math to those of the other people you talked to?

3. Would you say that you are more or less prone to math anxiety than other people? How do you account for the levels of math anxiety that you or others feel?

4. What personal or professional goals do you have that can be affected positively or negatively by your success as a math student? In what ways will your success affect these goals?

5. What personal qualities do you possess that may help you succeed as a math student? In what ways can you bring these qualities to bear when you are feeling frustrated or having difficulty?

Use the internet to find biographies of mathematicians by searching on the keywords "mathematics" and "biography" together. (The MacTutor History of Math site is probably the best.) Scan a few biographies until you find two, one of a man, one of a woman, that strike you as interesting. Make a note of the most interesting details of their lives, and include their birth and death dates and countries of origin. Compare your findings with those of your classmates.

Chapter 4

Strategies for Success

MATHEMATICS, as a field of study, has features that set it apart from almost all other scholastic disciplines. On the one hand, correctly manipulating the notation to calculate solutions is a skill, and as with any skill competence is achieved through practice. On the other hand, such skills are really only the surface of mathematics, for they are only marginally useful without an understanding of the concepts that underlie them. Consequently, the contemplation and comprehension of mathematical *ideas* must be our ultimate goal. Ideally, these two aspects of studying mathematics should be woven together at every point, complementing and enhancing one another, and in this respect studying mathematics is much more like studying, say, music or painting than it is like studying history or biology.

math is different

In view of mathematics' unique character, the successful student must devise a special set of strategies for accomplishing his or her goals, including strategies for lecture taking, homework,

39

and exams. We will examine each of these in turn. Keep in mind that these strategies are suggestions, not laws handed down from the mountain. Each student must find for him or herself the best way to implement these ideas, fitting them to his or her own unique learning styles. As the Greek said, know thyself!

4.1   Taking Lectures

Math teachers are a mixed bag, no question, and it's easy to criticize, especially when the criticism is justified. If your own math teacher really connects with you, really helps you understand, terrific—and be sure to let him or her know. But if not, there are a couple of things you will want to keep in mind.

the teacher's job

To begin with, think what the teacher's job entails. First, a textbook must be chosen, a syllabus prepared, and the material being taught (which your teacher may or may not have worked with in some time) completely mastered. This is before you ever step into class on that first day. Second, for every lecture the teacher gives there is at least an hour's preparation, writing down lecture notes, thinking about how best to present the material, and so on. This is on top of the time spent grading student work—which itself can be done only after the instructor works the exercises for him or herself. Finally, think about the anxiety you feel about speaking to an audience, and about your own math anxiety, and then imagine what a math teacher must do: manage both kinds of anxiety simultaneously. It would be wonderful if every instructor were a brilliant lecturer, but even the least brilliant deserves consideration for the difficulty of the job.

The second thing to keep in mind is that getting the most out of a lecture is *your* job. Many students suppose that writing furiously to get down everything the instructor puts on the board is the best they can do. Unfortunately, you cannot both write the details and focus on the ideas at the same time. Consequently, you will have to find a balance. Particularly if the instructor is lecturing from a set text, it may be that almost everything he or she puts on the board is in the text, so in effect it's written down for you already. In this case, make *some* note of the instructor's ideas and commentary and methods, but make understanding the lecture your primary focus. One of the best things you can do to enhance the value of a lecture is to review the relevant parts of the textbook before the lecture. Then your notes, instead of becoming yet another copy of information you paid for when you bought the book, can be an adjunct set of insights and commentary that will help you when it comes time to study on your own. If you don't read the text before the lecture, read it afterwards. The idea is to visit the material more than once.

the student's job

Finally, remember that your success is your instructor's success too. He or she wants you to achieve your goals. So develop a rapport with the instructor, letting him or her know when you are feeling lost and requesting help. Don't wait until after the lecture—raise your hand or your voice the minute the instructor begins to discuss an idea or procedure that you are unable to follow. Use any help labs or office hours that are available. If

you are determined to succeed and your instructor knows it, then he or she will be just as determined to help you.

4.2   Homework and Self-Study

There you are, just you and the textbook and maybe some lecture notes, alone in the glare of your desk lamp. It's a tense moment. Like most students, you turn to the exercises and see what happens. Pretty soon you are slogging away, turning frequently to the solutions in the back of the book to check whether you have a clue. If you're lucky it goes mostly smoothly, and you mark the problems that won't come right so that you can ask about them in class. If you're not so lucky you get bogged down, stuck on this problem or that, while the hours slide by like agonized glaciers, and you miss your favorite TV show, and you think of all the homework for your other classes that you haven't got to yet, and you begin to visualize burning your textbook...except that the stupid thing cost you ninety bucks....

Let's start over.

There you are, just you and the textbook and maybe some lecture notes, alone in the glare of your desk lamp. Relax. What are you here for? For whom are you doing this homework? Your teacher? Your parents? No, homework is for you, and you alone. It is your opportunity to learn, and to begin to gain mastery—and that is what you are here for. Not a grade—knowledge. Presumably your instructor has just lectured the material, but have you read the material in the textbook yourself yet? You haven't? Then do so. Reading the textbook is some-thing practically no student does, yet it can make a world of difference in how difficult the material seems to you. When reading a textbook, remember that it is not a novel, nor indeed like any other kind of book. Written math is *dense*. Each paragraph—sometimes even each line—contains deep ideas, which may require a novel way of thinking to un-derstand. It may take you 20 minutes or longer just to absorb and understand a single page. That is to-tally normal. Read it with blank paper available and a pencil in your hand. Work through the ex-amples yourself, until you thoroughly understand each step. Writing things down is far more effective than highlighting or underlining. Read the foot-notes. After you have done these things, then you are ready to look at the exercises.

reading math

Many instructors encourage their students to work together on homework problems. Modern learning theories emphasize the value of doing this, because students who collaborate can develop a synergy among themselves that supports their learning, helping them to learn more, more quickly, and more lastingly. Find out how your instructor feels about this, and if it is permitted find others in class who are interested in studying together. You will still want to put in plenty of time for self-study, but a couple of hours a week spent studying with others may be very valuable to you.

collaborating

4.3 Working Problems

As we noted earlier in this chapter, studying math is much more like studying piano or painting than it is like studying standard academic subjects. For imagine that someone who has never played piano sits down every day for a year and watches a concert pianist for one hour, as she plays works by the great composers for piano, such as Beethoven, Lizt, and Chopin. Finally, at the end of the year (having now watched hundreds of hours of the pianist's fingers striking the keys) this person sits down at the keyboard and tries to play. How will it go, do you think?

It won't, is the short answer, and it is easy to see why: although the study of music requires a considerable amount of rote learning of things like musical scales and keys, actually playing music is a skill, and like all skills you can't acquire it by watching someone else practice. Math is very similar. You can read the text and listen to the lectures, but until you start *doing* the math you haven't really begun to learn it yet. This is the great value of problem sets in a math class. They are, quite frankly, the best part of the whole enterprise. Simply put, math is not a spectator sport; the only way to get anything out of it is to get your hands dirty.

math is not a
spectator sport

Most problem sets are designed so that the first few problems are rote, and look just like the examples in the book. Gradually, they begin to stretch you a bit, testing your comprehension and your ability to synthesize ideas. Take them one at a time. If you get completely stuck on one, skip it for now. But come back to it. Give yourself time, for your subconscious mind will gradually formulate ideas about how to work the exercise, and it will present these notions to your conscious mind when it is ready.

As any experienced math instructor will tell you, some proportion of the students in any given class, on any given assignment, will look the exercises over and conclude that they don't know how to do it. They then tell themselves, "I can't do something I don't understand," and close the book. Consequence: no homework gets done.

Others will look the exercises over, decide that they pretty much get it, and tell themselves, "I don't need to do the homework, because I already understand it," and close the book. Consequence: no homework gets done.

Don't let this be you. If you've pretty much already got it, great. Now turn to the hard exercises (whether they were assigned or not), and test how thorough your understanding really is. If you are unable to do them with ease, then you need to go back to the more routine exercises and work on your skills. If, on the other hand, you feel you cannot do the homework because you don't understand it, then go back in the textbook to where you do understand, and work forward from there. Pick the easiest exercises, and work at them. Compare them to the examples. Work through the examples. Try doing the exercises the same way the examples were done. In short, work at it. You will learn mathematics this way—and in no other way.

If there is a problem you can't solve, then there is an easier problem you can solve: find it.

—George Pólya

Pólya's Problem Solving Advice

Mathematician and educator George Pólya wrote a now-famous book on studying mathematics titled *How To Solve It*, filled with outstandingly good advice on learning math generally and on solving math problems especially. We highly recommend the book—which is still in print nearly 70 years after it was written—but in lieu of that we here offer the highlights of Pólya's approach, which is organized by four principles.

First Principle: Understand the Problem.

It is surprisingly easy to bang your head against a problem without first facing up to this simple question: do you really understand what is being asked? After all, it hardly seems fair to expect yourself to answer a question you don't understand.

So, *before* attempting to solve a problem, get the answers to each of the following questions:

understanding the
problem

- Do you understand all the words used in stating the problem?

- What are you asked to find or show?

- Can you restate the problem in your own words?

- Can you think of a picture or diagram that might help you solve the problem?

- Is there enough information to enable you to find a solution?

Sometimes when we are struggling with a math problem, especially if we feel like we are *supposed* to understand it, we can neglect the need to ask the questions above to make sure we really do.

Second Principle: Devise a Plan.

There are many reasonable ways to solve problems. In fact, even when a good strategy has been chosen, there is usually another way (or several other ways) to attack the problem that would also be successful. Skill at choosing an appropriate strategy is acquired by solving problems—many of them.

Different types of math problems often call for different types of strategies, and reviewing particular strategies is usually an important part of your instructor's lesson plan for any given topic. Story problems are a special case—and an important one too, so we'll devote a whole section to them after reviewing Pólya's principles. In the meantime, here is a list, in no particular order, of very general strategies that it is always well to keep in mind:

strategies

- Guess and check.

- Make an orderly list.

- Eliminate possibilities.

- Use symmetry.

- Consider special cases.

- Use direct reasoning.

- Solve an equation.

- Look for a pattern.

- Draw a picture.

- Solve a simpler problem.

- Use a model.

- Work backwards.

- Use a formula.

A great discovery solves a great problem, but there is a grain of discovery in the solution of any problem. Your problem may be modest, but if it challenges your curiosity and brings into play your inventive faculties, and if you solve it by your own means, you may experience the tension and enjoy the triumph of discovery.

—George Pólya

Third Principle: Carry out the plan.

execution

Notice that this is the *third* principle. It is easy to just dive in working on a problem without attending to the first two principles. If you do that, this will be the hardest step. But if you are careful to deal with understanding the problem and carefully choose a plan first, then carrying out the plan is often the easiest step.

Sometimes, though, our plan doesn't seem to be working. Or we may, as we begin working, think of yet a better plan. There is no rule that says you should always stick with the plan you first choose,

flexibility

so don't be worried about stepping back, revisiting the first two principles (making sure you understand the problem and considering alternative approaches), and then starting afresh.

Remember, when you watch the instructor solve a problem s/he has prepared for class, you are seeing a performance already rehearsed. It can make it seem as though to them it's completely easy. What you are not seeing are the false starts and rethinking even the instructor often has to do when working a problem for the first time. So, never suppose that needing several attempts to crack a problem reflects negatively on your abilities; on the contrary, a willingness to try things that end up not working—and then to try a different approach—is the hallmark of those who ultimately succeed.

Fourth Principle: Look back.

When you have finished a problem—always a good feeling, of course—remember that you aren't really finished, not yet. Of course you should first satisfy yourself that the answer you got is the right one by checking it and ensuring it seems reasonable. (This is especially important with story problems, as we describe in the next section.) But then you should reflect on the process of finding the answer. What approach worked? Have you gained any insights into solving similar problems?

This process of reflection will help you enormously with the next problem you tackle, but perhaps more importantly it will cement in place the ideas and techniques you applied. This—this right here—*this* is learning mathematics.

review

4.4 Story Problems

Everybody complains about story problems, sometimes even the instructor. One is tempted to feel that math is hard enough without some sadist turning it into wordy, dense, hard-to-understand story problems. But again, ask yourself: "Why am I studying math? Is it so that I'll always know how to factor a quadratic equation?" Hardly. The study of math is meant to give you power over the real world. And the real world doesn't present you with textbook equations, it presents you with story problems. Your boss doesn't tell you to solve for x, he tells you, "We need a new supplier for flapdoodles. Bob's Flapdoodle Emporium wholesales them at $129 per gross, but charges $1.25 per ton per mile for shipping. Sally's Flapdoodle Express wholesales them at $143 per gross, but ships at a flat rate of $85 per ton. Figure out how each of these will impact our marginal cost, and report to me this afternoon."

the real world

49

The real world. In fact, successful students come to love story problems—because if you can work a story problem, you know you really understand the math. But it is important to have a good strategy and to apply it correctly. The worst mistake you can make—and the mistake most often made!—is to try to figure out how you are going to find the answer before you have completely analyzed the problem. To avoid this error, get in the habit of following the 7 Golden Rules of Story Problems:

7 Golden Rules of
story problems

1. Read the problem. Then read it again. There are two main things to focus on: the information given, and the question to be answered. So read it again and jot these down. (Just in case, read it again.)

2. DO NOT expect at this point that you will see how to do it. If it's a worthwhile problem, you definitely will not immediately see the way to solve it. So don't psych yourself out. After you have finished the next three steps will be soon enough to begin worrying about how you are going to find a solution. Honestly, it's better—and you have a better chance of solving the story problem—if at this stage you keep a completely open mind about what the solution will look like.

3. Next, note all the quantities described in the problem, write them down, and assign them labels. If the quantity is unknown, pick a letter such as a, b, P, M, etc. to represent it. This will permit you to work with the unknown quantity algebraically. (Avoid using x and y, since we are already trained to treat these in a

particular way, and these habits of mind may interfere with understanding the story problem.)

4. If appropriate, make a sketch of the situation described in the problem. Make it large and make it carefully, so that it accurately represents the situation. Then carefully label the sketch with all of the known and unknown quantities. Having a picture can help you understand the nature of the problem, and may even suggest a means of solving it. Remember, though, that a picture is not itself a solution, and in some cases it may even mislead you. Be careful!

5. Now focus on how the quantities (both known and unknown) are related to one another, and try to express these relationships in simple equations. It may be that certain formulas with which you are already familiar (such as equations for area, distance, etc.) will suggest ways of doing this. Remember the following English-to-Math translation tips:

translating English to math

a) "is" means "=".

b) "of" means "×" (times).

c) "per" means "÷" (divided by).

d) "proportional to" means "equals the same quantity multiplied by a constant." For example, "y is proportional to x" means "$y = kx$" for some constant k.

e) "inversely proportional to" means "1÷ (the same quantity multiplied by a constant)."

6. At this point, you want to focus on the equation that expresses the quantity you are trying to find in terms of the other known and unknown quantities. If your equation has more than one unknown, then your other equations (from step 5) may provide an opportunity to express some of these in terms of the others.

7. Finally, when you have an answer, check to ensure that your answer is sensible. If it claims that the son is older than the father, or that driving from point A to point B took "negative three" minutes, then something is wrong. Check through your reasoning and your algebra (and your arithmetic) to see if you can find the problem.

Other strategies that may sometimes work include referring back to a simpler but similar problem, looking for a pattern, making a table of values, or even guessing at the answer and checking to see if it works. Sometimes, if you think you know what the answer should be, you can work backwards from the answer to the solution. (Refer to Pólya's Problem Solving Advice in the previous section for more general strategies.)

All of these techniques can be useful sometimes, but there is no denying that there is a certain knack to doing story problems, and improvement only comes with practice. Just don't get stuck on one particular kind of solution, since different problems may require a different approach. Using the steps outlined above will help you to get every story problem under control right from the start.

4.5 Taking Exams

For many students, this is the very crucible of math anxiety. Math exams represent a do-or-die challenge that can inflame all one's doubts and frustrations. Also, any math anxiety you feel can be compounded by test anxiety, which is a different thing and should be dealt with separately. (If you suffer from test anxiety generally, you should discuss this with your instructor.) While it is frankly not possible to eliminate all the anxiety you may feel about math exams, there are some important techniques and strategies that will dramatically improve your test-taking experience. As you read these, think about how to apply each one on your next exam.

1. Don't cram. The brain is in many ways just like a muscle. It must be exercised regularly to be strong, and if you place too much stress on it then it won't function at its peak until it has had time to rest and recover. You wouldn't prepare for a big race by staying up and running all night. Instead, you would probably do a light workout, permit yourself some recreation such as seeing a movie or reading a book, and turn in early. The same principle applies here. If you have been studying regularly, you already know what you need to know, and if you have put off studying until now it is too late to do much about it. There is nothing you will gain in the few hours before the exam, desperately trying to absorb the material, that will make up for not being fresh and alert at exam time.

   don't cram

2. On exam day, have breakfast. The brain con-

   eat breakfast

53

sumes a surprisingly large number of calories, and if you haven't made available the nutrients it needs it will not work at full capacity. Get up early enough so that you can eat a proper meal (but not a huge one) at least two hours before the exam. This will ensure that your stomach has finished with the meal before your brain makes a demand on the blood supply.

read the exam
before you begin
working

3. When you get the exam, look it over thoroughly. Read each question, noting whether it has several parts and its overall weight in the exam. Begin working only after you have read every question. This way you will always have a sense of the exam as a whole. (Remember to look on the backs of pages.) If there are some questions that you feel you know immediately how to do, then do these first. (Students sometimes save the easiest ones for last because they are sure they can do them. This is a mistake. Save the hardest ones for last.)

expect anxiety

4. It is extremely common to get the exam, look at the questions, and feel that you can't work a single problem. Panic sets in. You see everyone else working, and become certain you are doomed. Some students will sit for an hour in this condition, ashamed to turn in a blank exam and leave early, but unable to calm down and begin thinking about the questions. This initial panic is so common (believe it or not, most of the other students taking the exam are having the same experience), that it's just as well to assume ahead

of time that this is what is going to happen. This gives you the same advantage as when the dentist alerts you that this may hurt a little. Since you've been warned, there's far less tendency to panic when it happens. So say to yourself, Well, I may as well relax because I expected this. Take a deep breath, let it out slowly. Do this a couple of times. Look for the question on the exam that most resembles what you know how to do, and begin poking it and prodding it and thinking about it to see what it is made of. Don't bother about the other students in the room—they've got their own problems. Before long your brain (remember, it's a muscle) will begin to unclench a bit, and some things will occur to you. You're on your way.

5. Math exams are usually timed—but remember, it's not a race! You don't want to dally, but don't rush yourself either. Work efficiently, being methodical and complete in your solutions. Box, circle, or underline your answers where appropriate. If you don't take time to make your work neat and ordered, then not only will the grader have trouble understanding what you've done, but you can actually confuse yourself—with disastrous results. If you get stuck on a problem, don't entangle yourself with it to the detriment of your overall score. After a few minutes, move on to the rest of the exam and come back to this one if you have time. And regardless of whether you have answered every question, give yourself at least two or three minutes at the end of

don't rush

55

the exam period to review your answers. The oops mistakes you find this way will surprise you, and fixing them is worth more to your score than trying to bang out something for that last, troublesome question.

6. In math, having the right answer is nice—but it doesn't pay the bills. SHOW YOUR WORK.

show all work

7. Finally, place things in perspective. Fear of the exam will make it seem like a much bigger deal than it really is, so remind yourself what it does not represent. It is not a test of your overall intelligence, of your worth as a person, or of your prospects for success in life. Your future happiness will not be determined by it. It is only a math test—it tests nothing about you except whether you understand certain concepts and possess the skills to implement them. You can't demonstrate your understanding and skills to their best advantage if you panic through making more of it than it is.

put it in
perspective

When you get the exam back, don't bury it or burn it or treat it like it doesn't exist—use it. Discover your mistakes and understand them thoroughly. After all, if you don't learn from your mistakes, you are likely to make them again.

review the graded
exam carefully

We have now finished the coping and strategies part of this booklet. Before we move on, remember: math anxiety affects all of us at one time or another, but for all of us it is a barrier we can overcome. We have examined the social and educational roots of math anxiety, some common math myths associated with it, and several techniques and strategies for managing it, but other things could be said, and other strategies are available which may help you with your own struggle with math. Talk to your instructor and to other students. With determination and a positive outlook—and a little help—you will accomplish things you once thought impossible.

4.6 Response: Thinking About Problem Solving

Each of the problems below demands a creative approach, the sort of approach that is most helpful on story problems. They are best worked on in groups of three or four students.

1. A local thief is going to attempt to take the treasure from his lord's castle. Unfortunately the castle is surrounded by a square-shaped moat that is 10 feet wide. Moreover, the moat is stocked with ravenous piranha—so no swimming. Looking about, the thief discovers two 9 1/2-foot planks behind some nearby bushes. How can he use them to get into the castle?

2. "I am the brother of the blind fiddler, but brothers I have none." How can this be?

3. Mary (a precocious 10 year old) has a secret club that meets in her tree-house in her back yard. Each meeting starts with every member present shaking hands—using the secret handshake—with every other member present. Suppose Mary and 7 of her fellow

members are present at a meeting. How many handshakes are exchanged?

4. You volunteer to take care of a neighbor's pets while she is away overnight on a business trip. There are 10 pets altogether; some are cats and some are dogs. You are given 56 small scoops-worth of pet nibbles with instructions that the cats are to get 5 scoops of nibbles each and the dogs 6 scoops of nibbles each. How many dogs does your neighbor have?

5. How many cuts does it take to divide a candy bar into 5 equal-size pieces? Into 6 equal-size pieces? Into 97 equal-size pieces?

6. How many ways can you make change for a $50 bill using only $5, $10, and $20 bills?

7. Bethany and Fred began digging holes at the same time. When Bethany digs 8 holes, Fred digs 5 holes. If Bethany is busy digging her 72nd hole, what hole is Fred on?

8. Watermelons were donated for a classroom picnic, but two were too ripe to be used. Three groups of four teachers cut up the remaining melons, and each teacher cut up five melons. How many watermelons were donated?

9. Suppose you are baking cookies using a very precise French recipe. The instructions say to bake the cookies for exactly 9 minutes, but you have only one 7-minute and one 4-minute hourglass on hand. How can you be sure your cookies won't be scorched or underdone?

Chapter 5

Your Future in Mathematics

IN this book we have focused a good deal on the role mathematics can play in your ultimate personal and professional success. You have probably, by now, thought of reasons why one or more math courses at the college level will benefit you. However, the fact is there is an entire academic department (at large universities sometimes two departments!) whose business it is to provide a full range of math courses to support the full range of career choices a student might make.

Since a course catalog can fail to detail the ways in which a given course might be useful to *your* academic goals, we provide this final piece as a kind of Hitchhikers' Guide chapter to the offerings and support you will probably find in *your* school's math department. Reading through it and then meeting with your academic advisor to discuss it can be one of the best things you do as a student to turbo-charge your academic engines.

5.1 The College Math Curriculum

This outline is not meant to take the place of advising, and you should not use it alone to plan your curriculum. **Always seek the advice of your assigned advisor** before making any registration decisions.

A typical college math department serves the needs of several different kinds of students. Many students only take mathematics as part of their "general education" or "distribution" requirements, and do not require any additional or specific mathematics courses for their major or career. However, most students will require some additional freshman or sophomore-level courses, such as statistics or calculus, as preparation for more advanced work within their chosen major. A smaller number of students require upper-division courses in logic, differential equations, and/or linear algebra for such majors as computer science and physics, and finally some students major in math itself, or statistics, and take 60 credit hours or more of mathematics coursework to qualify for their degree.

The following outline of a typical math department catalog is intended to acquaint you with the nature and purposes of the most common courses.

Gen-Ed Math (Freshman)

General Education courses come under many names; Math 100, College Math, Quantitative Reasoning, and Math for Liberal Arts, to name a few. These courses are intended to reinforce a basic set of skills at an adult level, and are often designed to interest the student in the applications of mathematics in their daily lives and/or to develop their appreciation of the aesthetic, philosophical, or cultural appeal of the subject. These courses typically do not count as credit towards any major that requires more advanced mathematics courses.

College Algebra (Freshman)

This course typically duplicates what would be covered in two years of high school algebra, and is a necessary preparation for all more advanced courses. It covers the number systems, basic algebra, equations and inequalities, polynomials and factoring, algebraic and transcendental functions, matrix algebra, analytic geometry, techniques of graphing, and often the use of appropriate kinds of technology.

Precalculus (Freshman)

Sometimes referred to as a "functions" or an "advanced algebra" course, precalculus extends the use of algebraic techniques to study and graph various kinds of functions, including polynomials, rational functions, exponential and logarithmic functions, and (if not taught in a separate course— see below) trigonometric functions. This course is specifically intended to prepare students to take calculus.

Trigonometry (Freshman)

If taught separately from precalculus, a course in trigonometry introduces the trigonometric functions, the trigonometric identities, graphing and applications of trigonometric functions, and sometimes spherical or hyperbolic trigonometric functions as well. This course assumes competence in college-level algebra, and is essential for taking calculus and other more advanced topics. Trigonometry is also a very appealing subject in its own right. It is largely self-contained, yet it provides insight into an astonishingly broad array of phenomena, such as the analysis of the interaction of forces in any physical model, the modeling of periodic systems such as heartbeats and planetary orbits, and the digitization of sound and other wave-like signals to make them available for computational treatment (so you can watch movies on your computer, for instance). Students see the world with fresh eyes after a course in trigonometry.

Calculus (Freshman/Sophomore)

The calculus cycle (or sequence), as it is called, typically runs over $1^{1}/_{2}$ to 2 academic years. Students in many majors will only require the first course ("calc 1"), which covers single-variable differential calculus, and may also introduce the definite integral. The second and/or third course covers additional techniques of integration, the use of sequences and series to represent functions, and sometimes also advanced analytic geometry. The third and/or fourth courses will cover differential and integral calculus of functions of two or more real variables. They may also cover calculus of vectors and of parametric functions. These latter courses are typically only taken by students in the mathematical, statistical, computational, or physical sciences, although they are accessible and can be very rewarding to any student who succeeds at the earlier courses. Calculus is the bridge from classical mathematics to the modern world. It is the essential set of insights that make possible the effective modeling of almost every kind of dynamic system, from the movements of the heavens to the pricing of manufactured goods. A sound grasp of calculus empowers the student to solve an enormous range of worldly problems.

Linear Algebra (Sophomore/Junior)

Linear algebra begins with the solving of systems of linear equations, but quickly takes students into their first taste of abstract spaces and generalized algebraic and geometrical insights. It is often taken concurrently with multivariable calculus. The techniques of linear algebra are important in almost every quantitative field, but especially in computer science, physics, business, and cryptography. Linear algebra also lays the groundwork for more advanced mathematical study of abstract spaces. Like trigonometry, linear algebra gives a whole new way of seeing and interacting with both the natural and the man-made world.

Statistics (Freshman–Senior)

Statistics is one of the main branches of applied mathematics, and in most departments forms a concentration within the major or, sometimes, even a distinct major all its own. Because the subject is fundamental to measuring almost every kind of phenomena, any student entering the physical, life, or social sciences, business, or economics will need at least a first course in statistics, and many will need more. Those who finish a degree in mathematics with a statistics concentration will study advanced statistical methods and experimental design, and can look forward to highly lucrative careers in industry as systems analysts or in business or government as actuaries (risk and insurance analysts).

Differential Equations (Junior)

Differential equations are equations in which function differentials (studied in calculus) are included as terms. The solution of a differential equation is a function, and being able to find such solutions is a central concern in every branch of physical science. It is not an exaggeration to say that serious physics begins with differential equations. This too is a gateway to entire fields of advanced study in mathematics.

Foundations (Sophomore/Junior/Senior)

At the sophomore or junior level this is a first course in symbolic logic, formal set theory, and the methods of proof used in mathematics. Courses of this type are sometimes referred to as *bridges* to abstract mathematics, because they teach the fundamental concepts and techniques used by all working mathematicians. This is also the student's first exposure to the nexus between mathematics and some related philosophical concerns—an enticing field of study. At the senior level students undertake an axiomatic approach to set-theory, including transfinite cardinalities and transfinite arithmetic, the non-denumerability of the real numbers, and the continuum hypothesis. Students would also expect to review the most important 20th century discoveries in mathematical logic, including the completeness and incompleteness theorems of Kurt Gödel.

Geometry (Freshman/Senior)

A freshman geometry course typically covers plane Euclidean geometry, including compass-and-straight-edge constructions and two-column proofs. Many programs no longer offer a geometry course at this level as it duplicates high school requirements in most states. An advanced geometry course will review the foundations of Euclidean geometry as a springboard to an axiomatic approach to non-Euclidean geometries. The study of non-Euclidean geometry is important for understanding modern physics because Einstein's relativity theory shows that the geometry of the actual universe is non-Euclidean. It is also important in pure mathematics both for its own sake and as a launch pad for the study of many kinds of abstract spaces.

Number Theory (Junior/Senior)

Number theory is at once one of the oldest and one of the most active current fields of mathematics. The techniques of number theory are of critical value in cryptography, and many number theorists are recruited by the government for this reason. Number theory is also fundamental to many fields of mathematics, from mathematical foundations to the most advanced analytic and algebraic fields. Many of the most important unsolved problems in mathematics (such as the Riemann Hypothesis) are problems in number theory. Among the most celebrated of recent accomplishments in the field is Andrew Wiles' proof of the 350-year-old "Last Theorem" of Pierre de Fermat in 1992.

Abstract Algebra (Junior/Senior)

When features and properties of the standard number systems are deeply understood, one may abstract their properties to more general mathematical structures. The most important of these structures are called groups, rings, modules, and fields, and they are the subjects of study in abstract algebra. This course is a rite of passage; it puts paid to all previous algebra study, and opens the door to the world of higher mathematics.

Analysis (Senior)

Usually called "intro to analysis" or "advanced calculus," this course does for the study of functions what abstract algebra does for the study of algebra—it takes students to the next level, where the properties of functions are studied in the abstract. Analysis and abstract algebra are the two most basic categories that organize all the myriad topics of modern mathematics.

Complex Variables (Senior)

Calculus is taught over the so-called field of real numbers. However, many branches of science and mathematics require a different field, called the field of complex numbers. In this course students learn how to apply everything they have learned in calculus to this new domain.

Numerical Analysis (Senior)

Advanced calculus is of only theoretical interest if it cannot be used to provide hard answers to practical questions. Numerical analysis is the study of the computational techniques and algorithms used to determine useful solutions. This course is the gateway to an enormous variety of pursuits known collectively as "applied mathematics."

Topology (Senior)

Sometimes called "rubber sheet geometry," this topic is intimately connected with analysis, and depending on how it is taught can be very abstract and set-theoretical. In this way it bridges the conceptual divide between mathematical foundations and the far reaches of calculus. An undergraduate course necessarily only scratches the surface of this deep and intriguing topic that is full of strange concepts and mind-bending structures.

Operations Research (Senior)

This course comprises important topics, including linear programming, game theory, and inventory theory, used to create and study mathematical models of business organizations and related systems. Students who take up this field have a highly remunerative career ahead of them in industry.

* * * * *

In addition, most departments will offer a sequence of courses designed to be especially helpful to prospective primary and secondary school teachers, including advanced coursework on basic math as well as methods classes for training in the techniques of teaching mathematics.

methods courses
for teachers

Aside from this standard fare, many special topics may be offered by any given department, depending on student and faculty interests. These may include courses on combinatorics, graph theory, solid geometry, projective geometry, probability theory, stochastic processes, Fourier analysis, functional analysis, history of math, philosophy of math, differential geometry, cryptology, dynamical systems, chaos and fractal geometry, wavelets, matrices, and the theory of computation, to name just a few. One of the joys of studying mathematics is the endless and stimulating variety of mental landscapes that it offers to the student who is eager to explore.

various other courses

5.2 The Mathematics Degree

Some students find that as they gain confidence in their math courses their interest in mathematics itself begins to grow, and they are naturally curious about the career opportunities that await the graduate with a major or minor in math.

The Math Minor

A minor in mathematics typically requires the full calculus sequence, differential equations, linear algebra, and one or two courses at the senior level. Many students find getting a minor in mathematics an especially attractive option if their major already has a significant math requirement. Biologists, chemists, physicists, economists, and computer scientists find that more advanced coursework in their chosen field becomes much easier to master when they have a stronger math background. More importantly, those students considering graduate work in these fields discover that a minor in math generally moves their application for graduate school to the top of the stack.

A minor in mathematics is also attractive to students working in philosophy, as many of the central problems in philosophy are closely related to issues in the foundations of mathematics, as we noted in Chapter 1. Here again, a minor in mathematics is a considerable benefit to any student considering applying to graduate school. (If you are working in a philosophy department, ask your professors about this.)

Students in design and visual arts sometimes pursue a minor in math as a way of deepening their cognitive and intuitive insights into space, pattern, symmetry, and relationship.

Future teachers know that a mathematics certification means a greatly increased chance of working in the district and at the school of their choice, and even those who entered primary school education in part to avoid math courses have learned that the mathematical knowledge expected of educators at that level is greater than they supposed, and growing.

Regardless of your likely or chosen major, the members of your college mathematics faculty stand ready to support your academic and professional success. Make sure to review your goals and requirements in mathematics with your assigned academic advisor.

The Math Major

So, finally, why would someone wish to major in mathematics? We hope by now you have come to see that mathematics isn't the province of "geeks and nerds," but a domain of knowledge that touches every aspect of our human experience. For many, from the ancient followers of Pythagoras to thousands of current researchers in universities and institutes around the globe, mathematics seems to be at the very heart of . . . well, everything. It is their passion and their reward.

But there are also quite practical reasons for studying mathematics, and many paths within the discipline to consider. People often suppose that most math majors usually go on to teach, but in fact fewer than one-third of students who obtain a B.A. or a B.S. in mathematics choose to remain in education at any level. The rest obtain employment in business, industry, and government. Statistics show that more than 90% of students who apply for non-academic jobs with a degree in mathematics are employed within 6 months of graduation, with an average starting salary above $53,000. Those who go on to obtain an M.A. or M.S. (typically a 2-year post-graduate degree) have average starting salaries above $75,000, and mathematicians with a Ph.D. generally command salaries in industry above $110,000, with no ceiling.

Careers for those with a bachelor's degree include actuarial science (insurance—very lucrative), systems analysis, professional services, quality control, operations management, accounting and finance, computer applications, and many others. Those with advanced degrees are typically in-

math degrees and business/industry

volved in research. The fact is, a graduate with a degree in mathematics is a proven problem solver, someone who can analyze a system or situation and determine quantitative solutions that work. Such a person is always highly valued by any business, industrial, or governmental enterprise.

Those who wish to remain in academia have opportunities as well. A bachelor's degree in mathematics, together with a teaching certification, qualifies you as a high school math teacher. You may have heard there aren't enough of these—such people are in very high demand. With a master's degree you can teach at a community college. The community college systems in almost every state are expanding rapidly as displaced workers find that they need to retrain, and as high school graduates increasingly turn to smaller, local colleges to give them an easier start at their college careers. The competition for academic positions at the Ph.D. level is stiffer, but the rewards are greater. Assistant, associate, and full professors of mathematics enjoy professional freedom, excellent working conditions, paid sabbaticals, top-of-the-line benefits, job security, and comfortable salaries (usually starting above $60,000 and going as high as $190,000 for tenured faculty at large universities).

math degrees and academia

There are other considerations that should encourage an apt student to look seriously at the math major. In most math departments the faculty/student ratio is the best on campus owing to the comparatively small number of students who choose to major in mathematics. Consequently, there is usually tremendous personal support available from faculty members who do not have to spread their time and attention among too many students. Small class sizes and a close-knit academic community mean a better educational experience, and far less risk that a student in difficulty won't get the help he or she needs to succeed.

You should feel free to approach your math instructor, or any member of the math department, for advice on how their program can help you achieve your goals.

5.3 Response: Thinking About Your Future

Write down three careers you believe you would find rewarding and that would meet your financial goals. For each of these three careers, list the probable math courses that would be required to get the appropriate degree. Also list any additional math courses that might support your goals in that career, either by helping you acquire the necessary knowledge and skills more easily or by making you more professionally competitive.

If possible, discuss these career choices with other students. Find someone who shares your goals, and compare notes.

Finally, make an appointment with your academic advisor to discuss the options that are available to you. Ask about prerequisites, related coursework, and the benefits of a related minor degree. If you find there is a math course that you wish to take, but feel nervous about, go to the math department and enquire at the office for someone who teaches it. Make an appointment with him or her, and discuss your reason for taking the course. Ask the instructor for advice on how to succeed.

Index

Platonic Realms

A Mathematics Encyclopedia

Articles and Minitexts on Popular Math Topics

A Database of Mathematical Quotation

Daily Humor, Quotes, and Math History Notes

Books, Posters, and Gifts

PlatonicRealms.com

www.ingramcontent.com/pod-product-compliance
Lightning Source LLC
Chambersburg PA
CBHW062107090426
42741CB00015B/3353